RICHARD DONOVAN GLOVER

TRACING OUR
GENETIC
MUTATIONS

FROM ME TO YOU

AuthorHouse™ UK
1663 Liberty Drive
Bloomington, IN 47403 USA
www.authorhouse.co.uk
Phone: 0800.197.4150

Published by AuthorHouse 07/17/2018

ISBN: 978-1-5462-9531-0 (sc)
978-1-5462-9530-3 (e)

authorHOUSE®

Contents

PREFACE

From a young age, I wanted to trace my family's history, hoping to get back to the very beginning of modern mankind – seemingly a ridiculous and impossible task! However, inspiration came to me in the 1950s whilst serving in the Royal Air Force. I was stationed for two years at Castel Benito in Libya, which was situated on a sandy, unmetalled road leading from Tripoli into the Sahara. Many times, it intrigued me to see people passing our camp on foot – some with shoes, some without, all with differing skin colours and facial features. I realised many of them must have walked hundreds or even thousands of miles across the desert towards Tripoli, hoping for a better life. A few of them obviously came from tropical Africa, with their ebony skin; some were the Tuaregs and other tribes of the desert, with their much-weathered dark-brown skin; and others were the Arabs of North Africa, with their lighter olive-brown skin, all in comparison to the former colonial Italians, with their tanned white skin. I was determined then to learn all about human migrations and the evolution that went along with them.

In the following pages, I've tried to explain these and other bodily differences and why they came about. These travellers along this sandy road descended from the people who had had various mutations many thousands of years earlier.

I've tried to bring this fascinating story to you in an easily readable form. The male Y chromosomes and the female X mitochondria in the cells of everyone alive today have travelled a 270,000-year journey, from the time of the first known anatomically modern humans – *Homo sapiens* – to today.

As Spencer Wells of the Geno 2.0 Project states, 'The greatest history book ever written is the one hidden in our DNA.'

1

MAP OF THE LONG JOURNEY OF
MY LITTLE Y CHROMOSOMES

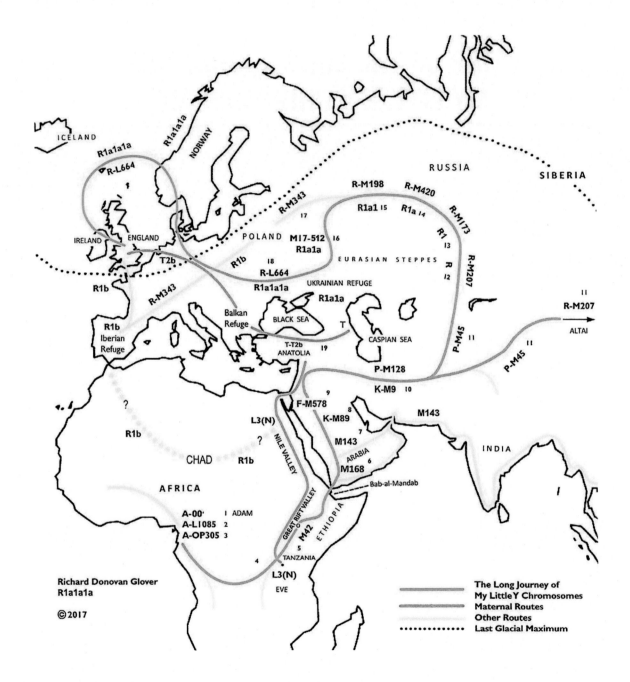

The long journey of mankind starts in Central West Africa and takes us to Ethiopia, the Great Rift Valley, Arabia, the Middle East, South-West Asia, Egypt, Levant, Southern Siberia, the Eurasian Steppe, Ukraine, Anatolia, the Balkans, Russia, Eastern Europe, Central Europe, Poland, Germany, Scandinavia, Norway, Ireland, and England.

I detail the small numerals on the map near each mutation number in the 'Haplogroups' section of this book. They give the possible origins and suggested dates of when each mutation took place. My ancestry, combining both my paternal and maternal sides, states that I am 43% Western/Central European, 34% Scandinavian, 19% British/Irish, 3% Jewish diaspora, and 1.3% Neanderthal. These percentages do not include African or Asian data. Overall, I am YDNA R-L664 (R1a1a1a) DYS388-10. On my maternal side, I am T2b. Interpreted another way, my ancestors in Europe must have lived nearly half their lives in Western and Central Europe, a third of their lives in Scandinavia, and only a fifth of their lives in the British Isles.

2

INTRODUCTION

I am not a scientist but a long-time and very interested spectator of human evolution and migrations. So I am not quite a DNA newbie. I leave the geneticists and scientists to do their work on this continuing, very complex, and difficult subject. I simply try to compare, assimilate, and translate information from the many published papers. I use this knowledge to explain my family history and, possibly inform many thousands of other people who know, or think, they have a Viking ancestor. Also, nearly everyone whose ancestors started their journey moving north to the Middle East and the Eurasian Steppe before dispersing east and west may find this book of interest. Varying mutations caused their characteristics to change to fit the climate they moved to.

I try to go back to the beginning of modern humankind, more than a quarter of a million years, when Homo's first major mutation took place to become Homo sapiens. But all the dates I mention are still speculative, and some are hypothetical. Various authors seem to differ in their conclusions. Until further genetic research is complete and archaeological remains are unearthed, it remains impossible to prove, without some doubt, this part of our evolution and migration. Like life itself, the climate at the time and overall climate change overall moulded these migrations.

We can see that the trail of Y male chromosomes and female mitochondria follows each step of the way in this discourse, from 270,000 years ago, A-00, to present day, R-L664. This is the journey I pursue today.

3

DNA AND ITS WORKINGS

Genetic mutations are permanent alterations in the DNA sequence that makes up genes. Gene sequences differ in most people by anywhere from a single DNA building block (base pair) to a large segment of a chromosome that can include multiple genes.

Our DNA is a long, linear sequence of small molecules called *nucleotides*. They constitute the double-helix DNA strand, denoted by the letters A, C, T, and G. They represent the building blocks of life. Their specific order and position in the genome, known as *markers*, encode the information within the DNA. Random mutations in the DNA sequence occur as copying mistakes during cell division. This can involve replacing a single nucleotide with another or deleting nucleotides. Mutations are rare, given the large size of the genome: about six billion nucleotides. Children can have only about fifty single-nucleotide mutations that distinguish them from their parents. These changes, going on for many thousands of years, gradually accumulate, enabling people to fit the climate and environment they live in.

Unlike other chromosomes, the Y chromosome carries only a few genes; most of it does not recombine. This little or no change makes the Y chromosome ideal for investigating human evolution and for medical and forensic genetics. The Y chromosome is self-determining. Half the sperm cells made by men develop into males upon fertilization; those without a Y chromosome (XX) develop into females. It follows that a man inherits his Y chromosome from his father, who gets it from his father, and so on, back through time for tens of thousands of years. It is the same for the female side, using the mitochondrial DNA that passes from mother to daughter, with little change for tens of thousands of years.

Random beneficial mutations in the Y chromosome are extremely rare, usually happening only after many thousands of years. But when they do, the offspring multiply, prosper, and spread. The seemingly sudden change from one mutation marker to another originated naturally in one individual. But by the time human remains are located, analysed, and dated, thousands of years could have passed. It is possible minor unharmful or neutral mutations can occur in any individual and then get passed down and eventually lead to genetic spread and diversity, essential in evolutionary terms but causing little change in perceived characteristics.

In contrast, harmful mutations in an individual, which are much more common, can cause people to die off quickly or in just a few generations. Unfortunately, we can see these all around us to this day. The healthiest in body and mind usually win and prosper. That is natural selection at work.

Glossary of Technical Terms

A *gene* is a unit of inheritance. It carries the DNA for controlling genetic activity, which transfers from parent to offspring.

There are twenty-three pairs of *chromosomes,* threadlike structures of deoxyribonucleic acid (DNA) and protein that occur in the nucleus of each and every living cell in the form of genes which supply the body's protein.

A *genome* is a complete set of genetic material that occurs in all cells of the body.

The male *Y chromosome* pairs with the female *X chromosome.* Females are born with two X chromosomes.

A *haplotype* is a group of genes inherited from a single parent.

A *haplogroup* is a group of similar haplotypes that share a common ancestor.

A *mutation* is the changing of a structure of nucleotides (A, C, T, and G pairs) in a gene, resulting in various forms, which get transmitted to future generations.

Mitochondria are the little organelles that convert carbon dioxide into oxygen for respiration.

4

CLIMATE CHANGE AND ADAM AND EVE

Historically, climate change is the reason why human migration ever took place. Changes in the Earth's orbit in a hundred-thousand-year cycle cause climate changes. Also, the changing of the tilt of the Earth's axis every forty-one thousand years, together with a slight wobble every twenty-three thousand years, causes climate changes. Why? Ask Milutin Milankovitch, a climate scientist from Serbia, or blame Jupiter and Saturn.

Fossils indicate that anatomically modern Homo sapiens probably evolved in the Great Rift Valley of East Africa. Geneticists agree that everyone alive today can trace their DNA and their ancestors to a region of South-Western Ethiopia, where Homo sapiens appeared in a 190,000-year-old fossil record near Kibish on the banks of the River Omo. However, scientists now know that the trail of the male Y mutations of modern man supposedly coalesced in a single man (A-00), who originated on the coast of sub-Saharan West Africa roughly 270,000 years ago (see section 7).

The descendants of this one man – *chromosomal or genetic Adam* – are thought to have migrated to the Great Rift Valley, where the fossils of so-called Adam and Eve were later discovered. The first known fossils of mitochondrial Eve were found to be about 180,000 years old. They belong to macro-haplogroup L, which lies at the root of the mitochondrial tree and represents one of the most ancestral lineages of all currently living humans. Adam and Eve and their descendants are the only known, real beginnings of humankind. They would both have been the only members of one small tribe of hunter-gatherers. Although the remains of Adam and Eve were found at different times, they most probably had the same antecedents in the same tribe.

A major beneficial mutation on Adam's Y chromosome, which obviously had developed in Adam's antecedents, gave him greater cognitive ability,[1] making him more intelligent and linguistically advanced. This characteristic enabled him to pass on his ideas to his family and others in the tribe. All of them benefitted from his knowledge, particularly in his tool-making skills and his added hunting prowess. Tribes that did not possess this knowledge died out.

Some scientists theorise that a diet of fish and its oil held the key to Adam's brain development. But better nutrition in general would have helped, too. Adam's male Y chromosomes were almost the same as those of every man living today. Their beneficial mutations have lasted on their very long journey to the present. Adam and Eve's mutations were advantageous. Those who did not have the same beneficial mutations were eliminated. Natural selection took over, and humankind prospered and progressed.

It seems obvious that beneficial mutations in one or more genes in the brains of primates gradually progressed for millions of years but have only been found in the above fossils, long before the outdated theory of Adam and Eve came along.

---

[1] A study by Dr Shelby Putt of the Stone Age Institute in the United States reveals that brain networks that might underlie a turning point in the human brain's evolution towards more humanlike intelligence came about 1.75 million years ago with the discovery of two-sided stone axes.

5

THE START OF THE LONG JOURNEY

When modern humans first ventured out of Africa, they left a genetic footprint still visible today. By mapping the frequency of genetic mutation markers, we can create a picture of when and where modern humans moved around the world. Some evidence says the first modern humans left Africa in the Sahara's wet phase, when the land was green and productive. They probably crossed the Bab-el-Mandeb Strait (the Gate of Tears), between modern Eritrea and Ethiopia, to modern Yemen in the Arabian Peninsula. Some tribes settled along the coast of Arabia, including in the United Arab Emirates (125,000 years ago) and Oman (106,000 years ago), when the land was green and fertile. Fossils of these modern humans dating back 115,000 years have also been found in modern-day Israel.

6

THE SAHARA AND THE NEANDERTHALS

The Sahara and Red Sea coastal route in the Arabian Peninsula was arid before 140,000 years ago and again after 115,000 years ago. This leaves a window of twenty-five thousand years for human migrations to have taken place.

Subsequently, the cooling and drying of the climate led to aridity again that lasted about seventy thousand years. About fifty thousand years ago, the ice sheets of Northern Europe began to melt, introducing a moister climate in Africa. The Sahara became habitable again then for more than five thousand years. That created a period of extensive vegetation, animal life, and human settlements. These windows acted as magnets that caused animals and, therefore, humans to move away from their roots.

As mentioned previously, after 115,000 years, the Sahara and the Arabian deserts grew dry again. The few modern humans who had moved north into modern-day Israel died out, supposedly overtaken after mingling with the Neanderthals.[2] I challenge this assumed theory.

With the deserts changing back to an arid phase, food sources would gradually disappear. Game animals would die off, and modern humans with them. Neanderthals, being a hardier and stronger species, and having been in Europe and the Near East for at least 350,000 years, would already have adapted to the vagaries of climate change better than modern humans. They would probably have moved north to more productive lands.

My own DNA consists of 1.3% Neanderthal, while the average person is 2.3% Neanderthal, according to the Genographic Project. This suggests that cross-breeding must have taken place several times in the average modern human's ancestry. Neanderthals probably lived together for many generations, not in Africa but in Europe and Western Asia.

---

[2] The Neanderthals who spread across Europe and Western Asia were the immediate ancestors of Homo sapiens. They were a tough, resourceful, and archaic human species, rugged enough to adapt to and survive in the Ice Ages in Europe. They died out about thirty thousand years ago, the last known ones living in Gibraltar and Southern Spain at about the same time as the artistic R1b people reached Western Europe.

Neanderthals separated from modern humans in Africa about 600,000 years ago. They had 99.5% of the same DNA as modern man, which is the reason why the Neanderthals in Europe and Western Asia had the capability to successfully cross-breed with modern humans. Some scientists have suggested that modern humans outcompeted them or killed them outright. But new genetic evidence suggests that our ancestors made love with their 'cousins', possibly starting with the L3(N) people in the Middle East, and because of this, the Neanderthal lineage disappeared into our much larger human population. We now know that the hybrid vigour of the Neanderthal added to our immune system, enabling our bodies to detect bacteria, fungi, and parasites. Neanderthals also added to our tolerance to the cold in climate change.

This seems to imply that 600,000 years ago, Homo sapiens was already around, although in the much more primitive Homo heidelbergensis form, from which Neanderthals, Denisovans, and modern humans all descended. Homo heidelbergensis developed from the 1.8-million-year-old Homo erectus. Discussion in 2017 covered a fossil found in Jebel Irhoud in Morocco, which is between 300,000 and 350,000 years old. Were they truly modern humans, or just another extinct branch of our human tree? Not all DNA from archaic humans was beneficial; most was detrimental, but over time, natural selection weeded it out. (This extract comes from *Quanta Magazine*, 'How Neanderthal DNA Helps Humanity'.)

7

THE DEEP-ANCESTRY HAPLOGROUPS: A-00, A-L1085, A-P305, AND M42

Haplogroup A-00, previously an unknown haplogroup, was recently found in the Y chromosome of a modern African American man; his ancestors were originally from sub-Saharan West Africa. They have an estimated age of about 270,000 years – considerably older than current estimates for the age of all anatomically modern humans. All men can trace their ancestors to this one man and his lineage. Other members of this haplogroup have been found in modern-day Western Cameroon in Central West Africa. A-00 therefore becomes the mutation of genetic or chromosomal Adam, and must for now be called the foundational haplogroup of all known paternal lineages of modern mankind. Many *subclades* of the A macro-haplogroup are found in various North-West and Central African populations (see Wikipedia, 'Human Y-Chromosome DNA Haplogroup').

Mutation A-L1085 is one of the subclades of A-00 relevant to this study. A-L1085 is thought to have originated 140,000 years ago, about the same time as some of the A-L1085 descendants crossed Ethiopia and the Great Rift Valley of East Africa. Ethiopia today is about 40% A-L1085 people. Others moved north to West Africa, and the highest frequency moved to Southern Sudan.

A-P305, a basal lineage of A-L1085, originated on the coast of modern-day Cameroon about one hundred thousand years ago. Some migrated to Southern Africa, and their lineage is frequent amongst the San people of Namibia. Other A-P305 descendants moved to the Great Rift Valley and are the paternal ancestors of 99.9% of all people living today (see section 4). A-P305 achieves its highest frequency in the modern-day Bakola pygmies of Southern Cameroon (8.3%). About eighty thousand years ago, one of the A-P305 descendants developed a new mutation, M42, which has modern-day frequencies in the Mbuti forest pygmies of Central Africa and possibly in the Hadza of Tanzania.

Cameroon Rainforest: Haplogroups A-00, A-L1085, and A-P305
Photo from Internet Free Photos.

8

MOUNT TOBA AND ITS EFFECTS ON CLIMATE CHANGE

The M42 descendants developed a further mutation, M168, and moved away from their roots in Ethiopia about seventy thousand years ago to start moving east. Research suggests that only small groups, perhaps ten thousand people altogether, left Africa between fifty thousand and one hundred thousand years ago. Why did some people move on? They simply followed their food source, which moved because of their tribe's expansion or the poor climatic conditions. Wherever the game animals went, they went too.

One of the main reasons why they moved on was the Mount Toba super-volcanic eruption that occurred seventy-four thousand years ago in Sumatra, which caused considerable climate change and certainly contributed to the lengthening of the Ice Age. That was the biggest eruption in more than four million years. The ash plume was four thousand times bigger than that created by the modern Mount St Helen's eruption. The ash plume, moving north and west, devastated much of South Asia as far as the Arabian Sea, covering the ground with 2,800 cubic kilometres of ash. The gases, including sulphur, circled the globe for six to ten years, killing off vegetation, animals, and people, who had already migrated as far as India in the first migration going east. Toba also deposited ash as far as North-East Africa and would have killed off much of the vegetation there.

It is now recognized that Mount Toba's ash travelled at least seven thousand kilometres, possibly reaching as far as Malawi at the southern end of the Great Rift Valley in South-East Africa. Toba would also have affected Ethiopia – perhaps not as badly as lands farther north and east, but certainly enough to make people want to move on. The Earth cooled as the gases and ash reflected the sun's heat, affecting rainfall and climates across the globe. Much of the world slipped dramatically into a cold portion of the Ice Age, which lasted one thousand years. Long 'nuclear winters' followed, decimating much of the world's vegetation. Mount Toba was a major cause of climate change and the migration of peoples, who had to move on – or starve. Some scientific observers remain undecided on the effect of Toba in East Africa, until further work is concluded. However, they believe that only ten thousand modern humans were left living in the whole world after the eruption – mankind lived on by a thread.

Modern humans passed through this genetic bottleneck, in which genome diversity was drastically reduced, but what followed was dramatic growth amongst dispersed populations that had already moved north and east in the first migration out of Africa. That soon prompted mankind to start migrations again, for what else could humans do but follow the game and herds?

9

MUTATIONS M168 AND M143

Seventy thousand years ago, after the lands had recovered from the effects of Toba, the M168 mutation saw our species leave its homeland in Ethiopia and move east to the Arabian Peninsula. The aridity of the desert was at its maximum, forcing the people to move on to more productive lands.

The M168 people searching for food or escaping from adverse conditions would have crossed the hot, dry Afar Depression and the Red Sea at the Bab-el-Mandeb Strait to modern Yemen in the Arabian Peninsula. Due to glaciation in both the Northern and Southern Hemispheres, the sea level was up to three hundred feet lower than at present, and the strait was much narrower, making it easier for very primitive log boats to cross. Seventy thousand years ago, rains along the coast of the Arabian Peninsula made the land fertile and very much like their African homeland, enabling animals and, therefore, people to move on ever north towards Asia, out of their familiar hunting grounds and into unexplored lands. The M168 descendants became the only lineage to survive outside Africa. Like all other markers in this study, the M168 marker continues to the present day.

About sixty thousand years ago, the M168 mutation was followed by mutation P-M143 on the coastline of Arabia. This was the first major migration to take place outside Africa. Some members moved north towards the grasslands of the Middle East, beachcombing along the coast of Arabia, but others in this second phase went east to India and Southern Asia, reaching Australia fifty thousand years ago, a journey which took ten millennia.

10

MATERNAL HAPLOGROUP L3(N): THE NILE ROUTE

The common ancestor of the maternal L3 people who had lived in the Great Rift Valley for more than one hundred thousand years gave rise about sixty-seven thousand years ago to the macro-haplogroup L3(N) (or N), which went on to populate the rest of the world. From this branch, all female lineages outside Africa arose. All living humans have this N gene in their cells.

Fifty thousand years ago, the ice sheets of Northern Europe began to melt, introducing a warmer and moister climate in Africa; parts of the Sahara became habitable. Some of the L3(N) tribes moved north to live, hunt, and gather on the plains and lower slopes on the wetter and more temperate north-western side of the Ethiopian Highlands. Any migrants would gradually move north down the Blue Nile Valley to the Nile basin and to the confluence with the White Nile, finding their prey in productive land along the river. The semi-arid conditions along the Nile allowed grasses and shrubs to grow with some trees, especially near groundwater sources, and on the floodplains. The vegetation and episodic rain pools enticed animals that adapted well to drier conditions to move in. Mankind had only to follow, with some humans settling along the way.[3] Some of the L3(N) people left Africa by this route and crossed the Sinai Peninsula from Egypt into Asia. They moved to the Eastern Mediterranean region and South-West Asia and mingled with the Neanderthals, passing on some of the Neanderthal genes to future generations of humankind. Sixty-thousand-year-old skeletons of modern men and Neanderthal fossils have been discovered together in caves in Israel.

---

[3] It would be interesting to find out if some of these people who settled along the Nile were the forebears of the 'ancient Egyptians'. Work on this is still very much ongoing, but it is already known that a complex intermediate hybrid mixture of genes from the Horn of Africa – A-P305, E-M78 (E-M35), and E-V12 – and sub-Saharan Africa plays a part.

Because much of the world's water was taken up by glaciation during the interpluvial, arid periods, the Sahara reverted to dry conditions again and expanded as the West African monsoon moved southwards. The flora and fauna retreated into the more fertile western Ethiopian Highlands and along the Nile valley. Any migrants at that time would have followed the animals and moved to disparate fertile areas. And eventually, they went along the Nile itself, with the help of crudely made log boats, to the Levant regions in the North. The nearly seven-thousand-kilometre journey down the full course of the Nile could have taken several thousand years, with the migrants living and hunting along the way.

The N people who lived alongside the M9, the M45, and the beginnings of the R mutation went along with them as they moved north and east into unexplored territory.

The Sahara Desert and Nile valley: Haplogroup L3(N)
Photo from Internet Free Photos.

11

THE MIDDLE EAST ROUTES: MUTATIONS M143, M89, M578, AND P-M128 (K-M9)

It seems obvious that mankind had several migrations well before the dates evidenced by geneticists and anthropologists. When the conditions were right, they moved on to find more productive lands. As some of mankind moved ever farther north, with less hot and humid temperatures and lower ultraviolet radiation, the black or brown melanin pigmentation in their skin lessened, enabling more sunlight to penetrate the skin. The lightening of the skin became obvious. The near-black skin in Central Africa and Ethiopia very gradually mutated to the much lighter olive-brown skin of the Middle East. That is Darwin's natural selection taking place, favouring the survival of the fittest.

The M143 mutation of M168 descent occurred in Arabia, with some tribes going east to India and Southern Asia and on to Australia and others going north to more productive lands. M143 gave rise to the M89 mutation and its subclade branches, which contain 90% of the world's existing non-African male population. The M89 mutation first appeared in an arid period, probably in the Arabian Peninsula, about fifty-five thousand years ago. Those humans followed the great herds of wild game for thousands of years, until the drying of the land made them move north to the Caucasus and the steppes of Central Asia.

The M578 mutation originated in South-West Asia about fifty thousand years ago, with some migrants settling in that region for several millennia, during a fertile wet phase which lasted five thousand years. Eventually, because of the drying of the land, they migrated in various directions, some towards Anatolia and the Caucasus and others to the Eurasian Steppe, together with some of the N people who had moved up with them.

M578 was followed about forty-seven thousand years ago by haplogroup P-M128, or K-M9, in South-West Asia, with the unique marker that descendants took with them for more than thirty thousand years. These descendants populated much of the planet. This marker is found in more than half of all non-Africans alive today. The M9 people moved with the game animals and herds, some settling and forming DNA clusters as they went, leaving their genetic trail behind them. The population expanded in these game-rich areas and then moved east, where cave dwellers in Northern Iraq left evidence of seeds, wild dates, acorns, and wall and pine nuts. Then, they possibly travelled over the Zagros Mountains to Iran and beyond, until they encountered the mountain ranges of the Hindu Kush, the Pamirs, and the Tien Shan. Some of these nomadic hunter-gatherer tribes stayed put, but others eventually dispersed in different directions – some north, some farther east to Siberia, and some south, crossing the high pass to India, China, and the Far East – each migration traced by its own genetic marker.

About forty thousand years ago, the climate changed again, becoming colder and more arid. Draught made the Middle East grasslands revert to desert once again, and the gateway from the South effectively closed to future migrations moving north.

The Arabian coastline: Haplogroups M143 and K-M89
Photo from Internet Free Photos.

12

SOUTH-WEST ASIAN ROUTE: MUTATION P-M45; EURASIAN ROUTES: MUTATIONS R-M207 (R), R-M173 (R1), R-M420 (R1A), R-M198 (R1A1), AND R-M17/M512 (R1A1A)

About thirty-five to forty thousand years ago, people with marker P-M45 branched off from M9 and turned north towards the game-rich steppes (treeless grasslands) of Kazakhstan, Uzbekistan, and Southern Siberia. Moving farther north again, they hunted buffalo, antelope, woolly mammoth, wild horse, bear, and reindeer. They armed themselves with spears, javelins, stone axes, and knowledge of fire. The M45 lineages survived on these Northern Steppe lands in the near-freezing[4] Ice Age climate because of their ability to make fire for cooking their food, for warmth, and for light.

The use of fire goes back many hundreds of thousands of years, first used by Homo erectus (more than 143,000 years ago). Without fire, the ability of mankind to move north into colder climes would have been near impossible. The absence of fire was also the reason earlier migrations always went to the warmer latitudes. The cave and hearth were the hub of flint- and tool-making activities and social interaction. It is suggested that the world population at that time was only in the tens of thousands.

About thirty-five thousand years ago, one man of M45 descent gained a new mutation – R, defined by marker and haplogroup R-M207. A small number of his descendants followed to the Northern Steppe, with its harsh conditions of sub-glacial tundra and arboreal forest. The R mutation ran alongside those of the ancient maternal N (L3) haplogroup, who were direct descendants of

---

[4] It is observed that inbreeding with the Neanderthals passed on genes that helped humankind adapt to cold climates.

mitochondrial Eve. The two groups had moved north and lived together for several millennia. However, some of the same R people – perhaps just a small band or family – diverged from the others and moved thousands of miles east to the Altai Mountains, which had long, brutally cold winters. Much earlier remains of our hominid cousins, Denisovans – another archaic cousin who seemed to mostly inhabit Asia – and the Neanderthals – who mostly inhabited Europe and Western Asia – have been found in Altai caves, preserved by the ever-present cold temperatures.

The Altai Mountains today
Photo from Internet Free Photos.

Evidence of the R mutation comes from the twenty-four-thousand-year-old remains of the 'Mal'ta boy' of the unique Mal'ta–Buret culture, who lived on the lower slopes or the plain below the high mountains of the Altai region, west of Lake Baikal, near modern-day Irkutsk. He belonged to a mammoth-hunting tribe that roamed across Siberia and parts of Europe during the Palaeolithic age. This one boy, and his ancestors a long time before him, had the very early R-YDNA chromosome marker, which contributed to the ancestry of most Europeans and South Asians.

The Mal'ta boy was a member of an area inhabited by a population with Mongoloid appearance.[5] They were a highly cultured and spiritual people who did not relocate to warmer areas but stayed in a small community. According to Russian archaeologist Mikhail Gerasimov's excavations in 1958, they lived in circular, semi-subterranean dwellings, with stone and clay foundations. Large animal bones, especially from woolly mammoths and woolly rhinoceroses, and reindeer antlers formed the walls. They covered their roofs in animal skins and earth sod to protect them from harsh, bitter northerly winds; they put a hearth at the centre.

The people of the Altai region could account for similarities amongst such spiritual systems as Chinese Taoism, Indian Vedism, and Tibetan Bon. The spread of many Asian languages, including Japanese, also stems from the Altaic. Full genome sequencing of the Mal'ta boy suggests a close ancestry may have originated in other Siberians and, after mixing with East Asians, made a substantial contribution to Native Americans of R1 (14–38%).

When the Great Ice Age, which started 120,000 years ago, began to lessen about 16,000 years ago, the glaciers in the Altai Mountains began to thaw, which forced the people to migrate to drier lands. Blocked by high mountains to the south and ice to the north, they moved east to the Himalayas, Eastern Asia, India, and China, taking their spiritual systems, their languages, and their Mongoloid appearance with them.

Amongst the accomplishments evident from the Mal'ta region are the remains of expertly carved bone sculptures and mammoth-ivory figurines of Venus-like forms and other works of art made of antler, similar but slightly different from the ones found in Western Europe. It therefore seems obvious the art of hand-carving must have been established in the Middle East, before the M45 and R people moved to Eurasia and Altai more than forty thousand years ago.

[5] The Mongoloid appearance of the R people in this small community was very different from the Proto-Caucasian or European features of the R people on the Eurasian Steppe. Their Mongoloid appearance was perhaps due to twenty thousand years of the expression of certain genes and natural selection working together, driving their bodies to rapidly respond to changing, stressful environmental conditions. Their present Mongoloid appearance was of straight black hair; dark-brown, almond-shaped eyes; and broad, flat faces. Everything adapted to fit the cold climate (Alekseev 1998, 323). This provides a fascinating example of the process of genomic differentiation in one group sharing a common ancestor. I believe the R mutation must have taken place in one man earlier than thirty-five thousand years ago, soon after or even before branching off from M9, as the M45 tribes moved north towards the Eurasian Steppe. After diverging from the main group, now with the R marker, the small band moved east, hunting their way towards the glacial Altai Mountains, taking many thousands of years to do so.

Between thirteen thousand and twenty thousand years ago, the main band of the R1, or R-M173, lineage had certain mutations showing characteristics of fair skin and light hair (so very different from the Mal'ta people). Beneficial mutations for blue eyes took place only six thousand to ten thousand years ago by reducing the melanin content in the iris (Human Genetics 2008). These characteristics developed, implying an origin close to Northern Europe, possibly the forest steppe of Northern Russia, with the genes acting under the stressful conditions of shortening daylight and the cold. These mutations worked to the advantage of the R1 people, assisted by natural selection, to correct the loss of the sun's ultraviolet radiation and the synthesis of vitamin D. Strong evidence says that both R1a and R1b people contributed to the mutation SLC24A5 (the A111T allele), which explains some of the 35% skin-tone difference between Europeans and Africans. The T2b Neolithic farmers in Anatolia carried this same R1 mutation from their ancestors, helping spread the lighter-skin mutation over most of Europe. The life expectancy of a hunter-gatherer, if the person lived into adulthood, was only about thirty years.

The R-M173 or R1 mutation appears to have originated before the Last Glacial Maximum, about twenty-four thousand years ago, and had several major levels of branching, which appear to have contributed to the ancestry of most Europeans. R-M420 (R1a) eventually colonized much of Poland, Western Asia, and modern-day Iran. R-M198 (R1a1) originated on the Eurasian Steppe, being the parent of the R-M17/R1a1a mutation, which is known as the oldest expansion out of the forest steppe. The R1b (R-M343)[6] mutation also originated on the Eurasian Steppe, became dominant in Western Europe about thirty thousand years ago, and reached Iberia about twenty-five thousand years ago. These were all R1a subclades.

---

[6] Some of the R1b people migrated back to Africa, possibly from Iberia and across the Straits of Gibraltar or from the Levant and Nile valleys, or even both, at different times. Ten thousand years ago, the Sahara was fertile and covered with forests and abundant animal life. It is now thought that some of the R1b (subclade V88) people – probably only small bands – moved south to fertile lands, reaching Chad between five thousand and seven thousand years ago, before the Sahara became arid once more. The R1b people then lived and mingled with the L3 and other tribes around the habitable Lake Chad Basin. Some of the Chadic people are still light-skinned; some have blue eyes but dark skin, generated from the mutations generated thousands of years earlier on the northern Eurasian Steppe. (For further reading, see *American Journal of Human Genetics: Cell Press*, volume 99, issue 6, pages 1316–1324.)

13

MATERNAL HAPLOGROUP T (T1, T2, AND T2B): THE FIRST FARMERS AND THE LINEAR POTTERY CULTURE

The maternal haplogroup T arose in a group of individuals who had a mutation which descended from one woman of the paternal R haplogroup, some of whom had genes for lighter skin and fair hair from mutations in the Northern Steppe thousands of years earlier. She lived about twenty-five thousand years ago in the Fertile Crescent of the Middle East, around Turkey (Anatolia) or Northern Syria and Northern Egypt. This one woman was my direct-line Stone Age clan mother.

Haplogroup T2 people originated in Anatolia about twenty thousand years ago. About ten thousand years ago, after the cold of the Younger Dryas (one of several interruptions to the gradual warming of the Earth's climate) had disappeared, their descendants' haplogroup T2b people became some of the first true farmers, who later migrated to Europe and spread their genes and knowledge. The climate and soils of Anatolia made the conditions perfect for developing and sowing tall grasses, including an early type of barley and primitive varieties of wheat, emmer, and einkorn. The climate made it possible to cultivate two sowings a year. They were already domesticating goats and sheep, making them some of the world's first agriculturalists. This farming technology enabled the people to settle and form permanent villages. They passed on their genes and particularly their knowledge to other groups. Other first farmers in the Zagros Mountains of Iran were a group distinct and different from those in Anatolia. Unlike the T2b, who migrated to Europe, they migrated to Asia.

As the ice fields of the Younger Dryas declined farther, many of the T2b people moved north to join the R1a1a, Kurgan, and R1b tribes on the Eurasian Steppe, spreading their technology and the Proto-Indo-European language with them. About 10% of native Northern Europeans now have the T2b haplotype. Today, T2b is found in high frequencies around the Alps and Southern Europe.

Anatolian grasslands: T1, T2, and T2b
Photo from Internet Free Photos.

In the second half of the fifth century AD, the Anglo-Saxon Belgae tribes invaded the South-East of England, bringing their Anglo-Frisian, or North Sea Germanic, Old English language and their culture, and some of them their T2b maternal genes with them. The British Isles today have about 5% of haplogroup T2b.

My maternal ancestors were known to have lived for many generations in Essex in the South-East of England, where the Anglo-Saxon raiders first landed. At that time, ordinary people were not allowed to move around the country. This, I believe, is where my maternal T2b genes came from: one of these Anglo-Saxon invaders, ultimately the first farmers in Anatolia ten thousand years ago.

14

THE EURASIAN STEPPE, THE KURGANS, AND THE PROTO-INDO-EUROPEAN LANGUAGE

About 18,500 years ago, my own haplogroup R1a1a, defined by the SNP marker R-M17, originated on the Southern Steppe, probably in the Ukraine region, and later became the most common male group in Europe. For many thousands of years, these people hunted in small groups, using traps, javelins, and bows and arrows. It is thought that these various Kurgan tribes could have colonized Poland fifteen thousand years ago. The people of both Poland and Ukraine today get 50% of their genes from R1a1a. The Kurgan culture is named for its method of burying its dead in barrows or mounds – a word originating in the Ural Mountains of Russia.

When the cold of the Younger Dryas began 12,900 years ago, R1a1a tribes moved south to the Black Sea and Caspian Sea (Caucasus) region of South Asia, sometimes called the Ukrainian or Rostov refuge. The people living in all other parts of Northern Europe also gradually moved south to refuges in Iberia, Italy, and the Balkans as cold weather advanced.[7] They had to let the animals lead the way. At that time, the British Isles were still connected to the European continent and covered with ice. The vast polar area was solid ice, and the Baltic Sea was a frozen lake. Ice covered most of Scandinavia, with a polar desert over most of Northern Eurasia. South of the ice, tundra covered most of

---

[7] About 11,600 years ago, after the Younger Dryas retreated north, some of the R1b people in the Iberian refuge spread along the western coast of France (the Pre-Celtic Atlantic Coastal culture). Some of them, after mixing with other groups, stayed in Northern Iberia. Thousands of years later, they became known as the Basques, who still speak the only isolated language with no known relationship to Proto-Indo-European or other modern languages. Some of the Basques, as direct descendants of the R1b hunter-gatherers, originated on the Eurasian Steppe. People from Wales have 92% of the basic R1b marker, people from Ireland have 81%, and people from the rest of Great Britain have more than 75%. Some of these people crossed over to the British Isles before the English Channel was made about eight thousand years ago.

Europe north of the refuges. South of the tundra lay dry steppe grassland, ideal for grazing animals.

About 11,600 years ago, the Younger Dryas ended very quickly, in only fifty to sixty years. After that, with the more temperate conditions of the Holocene Period, the R1a1a with the T2b people began to expand and move ever northwards to gradually repopulate the Eurasian landmass, living, hunting with dogs, and gathering on the Northern Steppe for many millennia. The steppes are considered the crucible of modern civilization (Anthony 2007).

The R1a1a Kurgan people of the Yamna, Corded Ware, Battle Axe, Pit Grave, and Barrow cultures of the Eurasian Bronze Age were found across the entire steppe region, from the Urals to Romania. Some of the Yamnas, primarily nomads, were now light-skinned and mostly brown-eyed. Some authors reported that an unknown 'ghost' population had been discovered, referred to as ANE (ancient Northern Eurasians). They were simply the Yamna Kurgans, who had had the necessary mutation thousands of years earlier.

About eight thousand years ago, the Kurgans met, mingled with, and learned from the first T2b farmers, going up the Danube valley from Anatolia with their farming technology. The Kurgans soon practised agriculture, animal husbandry, and the drinking of milk. Their mutated genes passed on lactose tolerance to many people in the world.[8]

Horse domestication is attributed to the Kurgans, sometime between 6,200 and 5,500 years ago. They also developed the use of early wheeled chariots and carts and the use of bronze weapons – distinct advantages over their enemies. They made and were protected by hill forts.

Some 4,600-year-old human remains in Eulau, Germany, were part of the Kurgan culture and were found to belong to haplogroup R1a1a. The Kurgan hypothesis is a composite of several of the above cultures and four successive periods and some of the first speakers of the Proto-Indo-European language.[9] During the Copper Age and Bronze Age, the Kurgans lived as nomadic pastoralists, but by the fifth millennium, they had expanded to Eastern

---

[8] Lactose intolerance occurs in only 5% of Northern Europeans but in more than 90% of people in some African countries, proving the genetic origins and spread of this mutation, which was thought to have originated on the Eurasian Steppe. However, we now know that lactose tolerance appeared in other world areas where people herded milk-producing animals. Lactose tolerance has arisen independently many times in the last nine thousand years.

[9] The Proto-Indo-European language was the world's most widely spoken language family – including English, the romance languages, Farsi, and various Indian tongues. A total of more than 445 languages and dialects are attributed to the Proto-Indo-European language, including North Germanic, which eventually formed the basis of the Nordic and Old Norse languages.

Europe, bringing their cultures and languages with them. The eastern area of Ukraine is the most convincing candidate for the origin of Indo-European languages.

The various Proto-Indo-European languages spoken by the Kurgan people spread westwards from the Eurasian Steppe by small bands of disconnected tribes, with their individual cultures imposing themselves onto local people as an elite. These migrations were not concerted military operations but expansions of disconnected tribes, eventually influencing and becoming most of the Balto-Slavic and related Eastern European languages, dependent upon where they settled.

Kurgan lands: The Northern Eurasian Steppe (R1, R1a, and R1a1)
Photo from Internet Free Photos.

The migrations lasted about two thousand years, originating from different parts of the Eurasian Steppe. Their various dialects and languages helped form the future country boundaries. This process was gradual and cultural, probably not a physical transformation, as was once thought, with hordes on horseback raiding westwards. They imposed an administrative system and religion on the indigenous group.

15

WESTWARD EXPANSIONS AND NORWAY: R-L664 (R 1A1A1A) AND THE EARLY BRONZE AGE

About 4,700 years ago, as the Ice Age retreated farther, these same R1a1a, Kurgan tribes, and T2b people – some with fair hair, blue eyes, and pale complexions – left their homeland on the steppes and Balkans, some now with an extremely rare haplogroup, R-L664 (R1a1a1a).[10] They gradually moved west through the Balkans and then north with other groups in their expansion of Yamna Corded Ware culture (named after cord-like impressions characteristic of their pottery). They travelled across sparsely populated Poland to what is now Germany, then Denmark, Sweden, and finally the fjords of Norway, reaching there in the Nordic Bronze Age, about 3,700 years ago. They left their genetic trail from the steppes behind them as they moved west and ever north, following the herds. Some of them took the new mutation and the Proto-Germanic/Nordic language with them. They went on a journey that took one thousand years, living, hunting, herding, fishing, and farming as they went. Other R1a1a and R-L664, with T2b, people moved west to North-West Europe, taking their Proto-Germanic language with them (see Eupedia's 'Haplogroup R1a (Y-DNA)' migration map).

My male genetic marker R-L664 is linked to Trondheim (the Trondelag region) on the coast of Central Norway, with a large genetic cluster of 31% of various R1a people living there today. This is where I believe my Viking

---

[10] Sir Francis Drake, the famous seafarer whom Queen Elizabeth I knighted in 1588, had the same rare male Y haplotype as I do: R-L664 (R1a1a1a), DYS388-10. Who was our shared ancestor? A Kurgan man from the Eurasian Steppe more than five thousand years ago. Sir Francis's ancestors probably came from the majority of the R-L664 people who moved to North-West Europe, rather than Norway.

ancestors came from before setting sail and marauding through other countries, including the British Isles.

Trondelag is located at the head of the broad and long Trondheimsfjorden, where 3,700 years ago, the land was nearly four metres lower than now due to the compaction of ice, making less land available for use. This fertile lowland around the fjord became the most important power centre of the Norwegian Viking age. Various cultures, including the R1a Yamna Corded Ware culture, made rock paintings there, proving beyond doubt that these same people migrated from the Eurasian Steppe thousands of years earlier.

Thirty-seven hundred years ago, ice still surrounded Central Norway much of the year, with the glacial mountains behind, but thanks to the Gulf Stream along the coast, the comparative warmth of this long, rugged strip of land enabled the people to grow cereals and vegetables in the summer months, much preserved and dried for long winter use. They worked with wood, bone, and flake implements, and some with bronze heads. They also had a high-protein diet of fish from the sea, salmon from the rivers, and seals from the coast – an ideal diet for maintaining and building bodily strength for future Viking warriors. Many of them still stayed mainly occupied with herding reindeer, as the Sami people do to this day, and hunted in the forests for aurochs, moose, European bison, and red deer. Many of these animals can be seen in rock carvings and Bronze Age petroglyphs in the Trondheim area (see Wikipedia, 'Rock Carvings in Central Norway').

As the Norwegian families expanded, they had to move on. They had no more land to work. With the seafaring skills possessed by their ancestors and passed down to them through more than two thousand years of sailing and boatbuilding, the sea and the major rivers became the only way forward.

The present-day Norwegian coastline
Photo from Internet Free Photos.

The Trondelag coastline: R1a1a, R-L664 (R1a1a1a), and I1a
Photo from Internet Free Photos.

16

THE VIKING PROPENSITY: THE BRITISH ISLES AND THE VIKINGS

It appears from the literature that the very same R1a subclades in Norway rarely met others farther south. But now they did! They met the people already there, the I1a, who had come up from their Balkan refuge with their darker hair, brown eyes, and darker complexions, and some of the R1b (M343) people, who had come up from their Iberian refuge thousands of years earlier. A few of the R1b people would still have the genes for lighter hair, lighter skin, and blue eyes, and some carried the gene for red hair from their Kurgan ancestors living on the Northern Steppe. More than 30% of Norwegians are I1a, adding to the mix of genotypes – just one of the reasons why Nordic people still have varying colourations from the mutations from many thousands of years earlier.

In a much smaller and congested area, up against the glacial mountains on one side and the often-frozen sea on the other, because of their Kurgan heritage, clashes with other people were inevitable. Fighting for the possession of land was necessary for their very existence. This posed a major problem for the Norse, especially after the eruptions of Icelandic volcanoes, which caused temperatures to lower for the next fourteen years. The extended darkness and cold was followed by the plague in the sixth century AD, when 75% of villages were abandoned. This disaster made the Vikings start their adventures of raiding and colonizing overseas. This was not a mutation but an epigenetic 'acquired characteristic' or trait, acquired from times long ago, which gets chemically incorporated or culturally maintained onto the genes. This warlike propensity can last for many generations until the environment changes for the better. It did, when they settled in other countries with more land available – but even then, they had to fight for it (see 'What You Don't Know About the Vikings' in the March 2017 issue of *National Geographic Magazine*).

One of the places the Vikings raided was, of course, the British Isles. The Norwegian Vikings went to Northumbria, the Scottish Islands, the Isle of Man (meaning Isle of Viking Sea Gods), and Ireland. The Danish Vikings mostly went to the East of England. My DNA chromosome signature states that I am of Norse Viking origin, haplogroup R1a1a (M17), and now R1a1a1a (R-L664), gained on the Eurasian Steppe. Today, about 5% of the people in the Western Isles, 2% in Shetland, 3% in Orkney, and 3% in other parts of the British Isles have Norwegian R1a genes. This R1a haplogroup does not include other Viking group data; for example, Shetland has about 30% and Orkney at least 20% of the total Norse Viking genotypes.

Norse Vikings invaded Ireland between the ninth and eleventh centuries, where they lived for more than a century. At first, the invaders faced a fierce resistance. Eventually, the centralized power of Norway declined, and local uprisings defeated the Norse. They were heavily defeated at the Battle of Tara in 980, which destroyed the entire army of more than two thousand Vikings, and the Battle of Clontarf near Dublin in 1014. After their defeat at Clontarf, the Vikings were expelled from the whole of Ireland and were forced to move to the Isle of Man, the Wirral, West Lancashire, and the Pennines. The last location ties in with the movement of one man called Manannain,[11] who gave his name to the settlement of Manningham (the home of Manannain) near Bradford, West Yorkshire. The settlement of Manningham was first mentioned in 1240. After that, the 'de Mayningham' or 'de Manningham' name appears many times in the history of Bradford, Yorkshire, and much farther afield.

Manningham is only about a twenty-five-mile straight line from the Ribble Estuary, where many of the expelled Vikings landed, and equally distant from the main Viking city of Jorvik (York). The possible journey from the Ribble would have been an easy one – up the Ribble valley, where the Viking Cuerdale Hoard was found near the banks of the river.[12] Then one would travel through the Pendle and Trawden forests, over to the eastern slopes of the Pennines to the River Worth, a tributary of the major River Aire. Manningham became the main route and a possible trading post for travellers between Ireland, the Isle of Man, and York.

---

[11] This is the Anglicised form of the Gaelic O'Murnain. This name was born in Celtic mythology for a sea god associated with the Isle of Man, a descendant of Mannanan.

[12] The Cuerdale Hoard dated from soon after the expulsions from Dublin. It was the largest Viking silver hoard found outside Russia, with 8,600 items of silver and gold, with many coins of hack silver made in Ireland (see *The Viking Hoard* by Williams and Webster, 2011, and 'The Rise of Viking Dublin' in *Current Archaeology*, issue 328, July 2017).

The Manningham family became the lords of the manor of Manningham, where, in 1298, John de Manningham is said to have engaged in a fight with Henry, the servant of John de Soothill and, in true Viking tradition, cut off his arm. In 1363, the Manninghams bought the rights to the lordship of the manor of Wrenthorpe, near Wakefield, which adjoined the manor of Soothill. King Edward IV knighted two members of the family, John and Oliver Manningham, after they fought in the fifteenth-century War of the Roses. This fighting legacy followed for many generations. I believe this tendency remains alive today – but only in fighting the hard slog of business (see *The Descendants of the Sea Gods: The Glover/Manningham Clan*, 2008). The Vikings of this family are my direct ancestors, and still further back in time are, without doubt, Kurgan warriors. What a legacy! They evolved with this belligerent acquired characteristic, which was always with R1a people, from more than twenty thousand years ago in the Eurasian Steppe.

The Manningham surname changed to Glover after the sixteenth-century Reformation, when the family lost the rights to farming the lands at Newlands Priory in Normanton, near Wakefield. They became glove makers, and hence the name change to Glover. The Manningham/Glover families have since expanded not only in the United Kingdom but to various other countries, notably the United States of America, Canada, Australia, and New Zealand. They remain there, establishing their businesses and farms – all from one family of ancestors more than 270,000 years ago.

CONCLUSION

The whole basis of human evolution and the migratory path of my genes from Africa to England revolve around climate change, natural selection, and the survival of the fittest. Without these, my genetic forebears would still live in the treetops and savannah grasslands of Africa. Evolution is a progressive march from primal origins to adapt and fit the terrain we live in.

We have seen why our skin changed from almost black to brown and then light tones. Our hair changed too, and some of our eyes changed from the dominant brown to blue – all because of climate change – as we moved north. Without climate change and natural selection, mankind would never have progressed. We would have had no need to adapt to changing conditions; we would have had no need to leave the sanctuary of the forest.

My little Y chromosomes together with my mitochondria have gone through quite a lot, from when the first modern man (Homo sapiens) lived more than 270,000 years ago in Central West Africa. So-called Adam and Eve became some of the first ancestors of everyone living today. My Y chromosomes multiplied by many billions in all my 12,000 or more generations. It took only fifteen known beneficial mutations to reach the present. All others fell away.

We can see how we lived and bred alongside our more ancient 'cousins', the Neanderthals, who became part of most of us. We can see how the effects of volcanic activity nearly annihilated all mankind. We can also see how one branch of my ancestors, more than 20,000 years ago, adapted to the cold climate of Siberia by changing their whole genome, altering their bodily features to better fit the climate and terrain they lived in.

Migrating from sub-Saharan West Africa more than 140,000 years ago to Ethiopia, my ancestors left Africa and survived both enemies and predators through deserts, forests, savannah, tundra, and ice. Across rivers and mountains, they moved to live and hunt on the Eurasian Steppe for more than thirty thousand years. As the ice retreated, my little Y chromosomes in one of the branches of the R1a1a1a Kurgan ancestors moved west and north in the Neolithic Expansion to Norway, living there for more than two thousand years as a Viking. More than a millennium ago, he moved to Ireland and England, always leaving his genetic trail behind, allowing science to virtually follow him all the way from Africa to England.

Twenty-five thousand years ago, my maternal T2b mitochondrial genes followed a similar journey from Ethiopia to Anatolia. Ten thousand years ago, people carrying these genes became some of the first farmers. They brought their knowledge and language and settled in most of Europe, staying in Northern Europe for more than a millennium. As the so-called Anglo-Saxons, they invaded and colonized much of Eastern England.

ACKNOWLEDGEMENTS

I would like to thank the following people and organisations. Thanks to Google, Wikipedia, and Eupedia, for without them, my study would certainly not be here at all. Thank you to National Geographic for their 2008 Genographic Project and 2016 Geno 2.0 Project (The Human Journey: Migration Routes) and for testing my DNA to show my haplogroup is R1a (M17), and the latest mutation, R-L664 (R1a1a1a). I thank Professor Bryan Sykes, professor of genetics at Oxford University, for testing my DNA and showing my Y chromosome signature to be of probable Norse Viking origin. I thank the Wellcome Trust and the University of Leicester for inviting me to take part in their People of the British Isles Project, which again showed my predicted haplogroup was R1a; also, one rare marker, DYS388-10, repeats, rather than the normal twelve. I would also like to thank Professor Adrian Wood of Huddersfield University for his help and encouragement. And thank you to Stuart Hartley for his help.

NOTES

Full references are not denoted in the text but come from the following articles and papers.

Human Origins, *National Geographic*, 1/2. University of Utah, Hauri yDNA Project, 1/6–2/6. 'Wikipedia: Early Human Migrations', Wikimedia Foundation, last modified October 1, 2017, 19:40, https://en.wikipedia.org/wiki/Early_human_migrations.

'Wikipedia: Early Human Migrations', Wikimedia Foundation, last modified October 1, 2017, 19:40, https://en.wikipedia.org/wiki/Early_human_migrations, 2/11.

National Oceanic and Atmospheric Administration, 'Climate Timeline Tool', accessed October 16, 2017, https://www.ncdc.noaa.gov/data-access/paleoclimatology-data, 2/3.

Jonathan Adams, *Africa during the Last 150,000 Years* (Oak Ridge, TN: Oak Ridge National Lab).

Human Journey, *National Geographic*, 1/2. Ice Age Civilization, 2/16.

Ice Age Civilization map, 9/16. 'Wikipedia: Sahara Pump Theory', Wikimedia Foundation, last modified September 20, 2017, 1:28, https://en.wikipedia.org/wiki/Sahara_pump_theory, 2/4.

Ancestry DNA, Ice Age, 1/9. 'Haplogroup R1a (Y-DNA)', Eupedia, last modified October 2017, https://www.eupedia.com/europe/Haplogroup_R1a_Y-DNA.shtml. Map. *National Geographic*, 'Genographic Project'. University of Utah, 'Hauri yDNA Project', 1/2–3/6.

National Oceanic and Atmospheric Administration, 'Climate Timeline Tool', accessed October 16, 2017, https://www.ncdc.noaa.gov/data-access/paleoclimatology-data, 1/3. Welcome to Adobe GoLive 4.

Human bottleneck: Richard G. Roberts, Michael Storey, and Michael Haslam, 'Toba Super Eruption: Age and Impact on East African Ecosystems', *Proceedings of the National Academy of Sciences of the United States of America* 110, no. 33 (August 2013): E3047. 'Wikipedia: Early Human Migrations', Wikimedia Foundation, last modified October 1, 2017, 19:40, https://en.wikipedia.org/wiki/Early_human_migrations, 3/11.

National Oceanic and Atmospheric Administration, 'Climate Timeline Tool', accessed October 16, 2017, https://www.ncdc.noaa.gov/data access/paleoclimatology-data, 1/3. The Toba Super Volcano and Human Evolution, 1/2.

'Wikipedia: Early Human Migrations', Wikimedia Foundation, last modified October 1, 2017, 19:40, https://en.wikipedia.org/wiki/Early_human_migrations, 4/11.

Climate improved seventy thousand years ago: 'Wikipedia: Early Human Migrations', Wikimedia Foundation, last modified October 1, 2017, 19:40, https://en.wikipedia.org/wiki/Early_human_migrations, 4/11.

L3 (M168) people: 'Wikipedia: Early Human Migrations', Wikimedia Foundation, last modified October 1, 2017, 19:40, https://en.wikipedia.org/wiki/Early_human_migrations, 4/11.

Eighty-five metres (three hundred feet) lower due to water in glaciers: Ice Age Civilization, Wikipedia, 8/16.

Wet phase: 'Wikipedia: Sahara Pump Theory', Wikimedia Foundation, last modified September 20, 2017, 1:28, https://en.wikipedia.org/wiki/Sahara_pump_theory, 1/4.

'Wikipedia: Sahara Pump Theory', Wikimedia Foundation, last modified September 20, 2017, 1:28, https://en.wikipedia.org/wiki/Sahara_pump_theory, 2/4.

'Wikipedia: Human Skin Color', Wikimedia Foundation, last modified October 9, 2017, 0:16, https://en.wikipedia.org/wiki/Human_skin_color, 1/22 and 2/22.

'Wikipedia: Sahara Pump Theory', Wikimedia Foundation, last modified September 20, 2017, 1:28, https://en.wikipedia.org/wiki/Sahara_pump_theory, refs. 6 3/4.

Climate timeline: Summary of one hundred thousand years. 'Wikipedia: Haplogroup F-M89', Wikimedia Foundation, last modified September 7, 2017, 7:19, https://en.wikipedia.org/wiki/Haplogroup_F-M89, 1/4.

Geno1 Map, *National Geographic*.

'Wikipedia: Mal'ta–Buret Culture', Wikimedia Foundation, last modified September 9, 2017, 13:21, https://en.wikipedia.org/wiki/Mal%27ta%E2%80%93Buret%27_culture.

Ancestry DNA, Ice Age, 1/9. 'Haplogroup R1a (Y-DNA)', Eupedia, last modified October 2017, https://www.eupedia.com/europe/Haplogroup_R1a_Y-DNA.shtml, 1/19. Ice Age Civilization, 8,9,10/16.

'Haplogroup R1a (Y-DNA)', Eupedia, last modified October 2017, https://www.eupedia.com/europe/Haplogroup_R1a_Y-DNA.shtml, 1/19 and 2/19

'Haplogroup R1a (Y-DNA)', Eupedia, last modified October 2017, https://www.eupedia.com/europe/Haplogroup_R1a_Y-DNA.shtml, 3/19.

'Haplogroup R1a (Y-DNA)', Eupedia, last modified October 2017, https://www.eupedia.com/europe/Haplogroup_R1a_Y-DNA.shtml, 9/19.

'Wikipedia: Kurgan Hypothesis', Wikimedia Foundation, last modified September 27, 2017, 19:15, https://en.wikipedia.org/wiki/Kurgan_hypothesis, 7/12, horse 9/12.

'Haplogroup R1a (Y-DNA)', Eupedia, last modified October 2017, https://www.eupedia.com/europe/Haplogroup_R1a_Y-DNA.shtml, 6/19. 'Wikipedia: Kurgan Hypothesis', Wikimedia Foundation, last modified September 27, 2017, 19:15, https://en.wikipedia.org/wiki/Kurgan_hypothesis, 2/12–5/12.

'Haplogroup R1a (Y-DNA)', Eupedia, last modified October 2017, https://www.eupedia.com/europe/Haplogroup_R1a_Y-DNA.shtml. 'Wikipedia: Kurgan Hypothesis', Wikimedia Foundation, last modified September 27, 2017, 19:15, https://en.wikipedia.org/wiki/Kurgan_hypothesis, 1/12–4/12.

'Wikipedia: Kurgan Hypothesis', Wikimedia Foundation, last modified September 27, 2017, 19:15, https://en.wikipedia.org/wiki/Kurgan_hypothesis, 6/12–7/12.

'Wikipedia: Kurgan Hypothesis', Wikimedia Foundation, last modified September 27, 2017, 19:15, https://en.wikipedia.org/wiki/Kurgan_hypothesis, 5/12.

'Wikipedia: Red Hair', Wikimedia Foundation, last modified October 14, 2017, 13:15, https://en.wikipedia.org/wiki/Red_hair, 1/15.

Last glacial maximum: 'Wikipedia: Haplogroup R1a', Wikimedia Foundation, last modified October 14, 2017, 3:22, https://en.wikipedia.org/wiki/Haplogroup_R1a, 1/6.

Ice Age Poland: 1/9 and 2/9. 'Wikipedia: Stone-Age Poland', Wikimedia Foundation, last modified July 2, 2017, 2:56, https://en.wikipedia.org/wiki/Stone-Age_Poland, 1/2–3/9.

Genographic Project, *National Geographic*.

Glover, 'Epigenetic Acquired Characteristics,' 2008.

'Genetic Maps of Europe', Eupedia, https://www.eupedia.com/europe/genetic_maps_of_europe.shtml.

HAPLOGROUPS

The suggested dates and origins listed here are taken from the text. See the small numerals marked on the map 'Map of the Long Journey of My Little Y Chromosomes' (BP = *Before Present*).

1. A-00. 270,000 BP (Adam) – Central West Africa
2. A-L1085. 140,000 BP – Central West Africa
3. A-P305. 100,000 BP – Cameroon, Africa?
4. L3(N). 180,000 BP (Eve) – Great Rift Valley, Africa
5. F-M42. 80,000 BP – Ethiopia
6. F-M168. 75,000 BP – Ethiopia
7. P-M143. 60,000 BP – Arabia
8. K-M89. 55,000 BP – Arabia
9. F-M578. 50,000 BP – Levant
10. P-M128–K-M9. 45,000 BP – Middle East
11. P-M45. 35,000–40,000 BP – Southern Eurasian Steppe
12. R-M207–R. 35,000 BP – Northern Steppe and Southern Siberia (Altai)
13. R-M173–R1. Before 18,500 BP – Northern Eurasian Steppe
14. R-M420–R1a. 30,000–35,000 BP – Eurasian Steppe
15. R-M459–M198, R1a1. 18,500 BP – Eurasian Steppe
16. M17–R1a1a. 18,500 BP – Eurasian Steppe
17. R-M343–R1b. 30,000 BP – Eurasian Steppe
18. R-L664–R1a1a1a. 4,700 BP – Eurasian Steppe
19. 19. T–T2b. 25–10,000 BP – Anatolia/Levant/South-West Asia

My R1a1a1a R-L664 short tandem repeats (STRs) are as follows: DYS 393–**13** 19–**15** 391–**10**
439–**10** 389i**14** 389ii.i–**16 388–10** 390–**25** 425–**12** 426–**12**
385a–**11** 385b–**14** 392–**11** 635–**23** 437–**14** 438–**11** 448–**19** 456–**15** 458–**15** YGATAH4–**12**.

The Author
Richard Donovan Glover
rg35416754@gmail.com

Printed in the United States
By Bookmasters

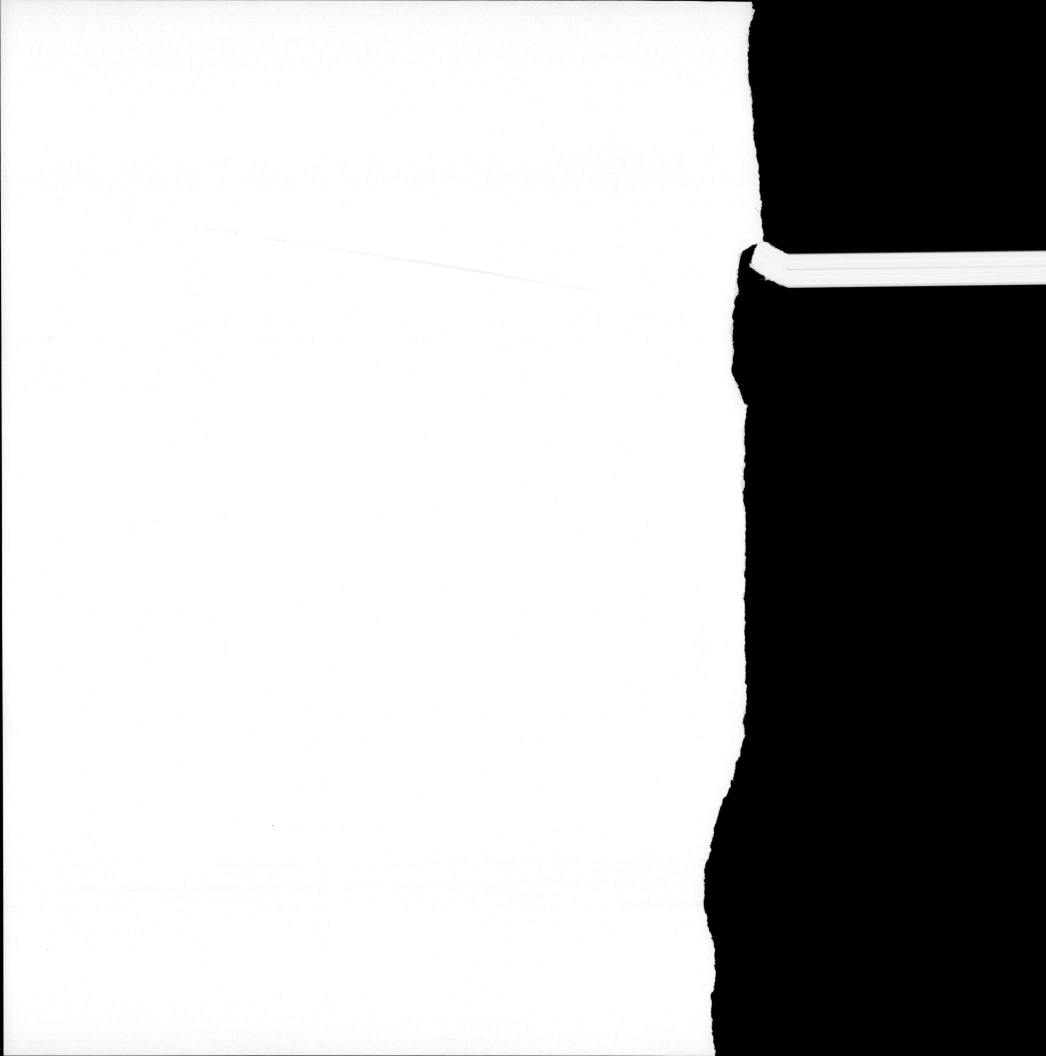